Biopaint- Preparation and Testing

Darani Vasudevan

Preface to the book

This book has details on the preparation of Biopaint from the sources available in nature and testing its effect on Algal growth in the walls of a temple in Erode District, Tamil Nadu, India. This was carried out as my research work and the dissertation was submitted to the Periyar University, Salem, Tamil Nadu, India in the year 2016. Hope this book will display an idea for the choice of Biopaint in real life applications.

-V.Darani M.Sc., M.Phil., (SET)

Contents inside....

Introduction

Materials used and the Methodology

Results

Discussion

Summary and conclusion

Bibliography

Biopaint- Preparation and Testing

Chapter 1: Introduction

A liquid or a mastic composition applied on any surface as a skinny layer either to impart colour or protection is termed as paint. Paint is in long term usage in the history of mankind for preventing corrosion. The art of painting was known since prehistoric periods where people used red ochre, yellow ochre, hematite, manganese oxide and charcoal for cave paintings. Paint is mostly used in liquid form but it is also available in the form of powders. The main ingredients in paints are pigments, binder, solvent or diluents, fillers and additives. Additives are mostly avoided while preparing biopaints.

The granular solids or colour giving compounds added to the paint are termed pigments. Natural pigments include plant dye, clays etc. The film forming component of paint is termed binder. It gives gloss, durability, flexibility and toughness to the paint and these properties varies depending upon the binders used. The main purpose of using solvents in paint is to adjust the viscosity of the paint and to control the flow and application properties of the paint. Fillers are those compounds that are used to increase the volume or bulk of the paint. Cheap and inert materials like diatomaceous earth, talc, lime stone powder, barites, and clay are frequently used as fillers in paint. The miscellaneous substances present in the paint apart from those mentioned above fall under the category additives. The additives are mainly used in small quantities to improve flow properties, control foaming and they also act as biocides.

The synthetic paints are widely used throughout the world. They have many useful impacts at the same time they cause adverse side effects to the humans and environment because,

- ➢ Most toxic solvents like toluene, xylene are used.
- ➢ Some synthetic pigments are carcinogenic.
- ➢ Many additives are toxins.
- ➢ Cause breathing problems and allergies.

➤ These synthetic paints are not eco- friendly.

The solution to avoid all these adverse effects would ultimately the use of Biopaint. Many plants yield a variety of dyes which are widely used in health, nutrition and various other fields. Plant pigments are relatively stable and could offer significant opportunity as a colorant in paints. Moreover biopaints are eco friendly paints. Organic materials used in these paints include natural pigments of minerals, plant and animal origin and other raw biodegradable ingredients (Nithyananda sastry *et al.,* 2016). The biopaints are durable and enhance a positive room climate. For the preparation of biopaints safer technologies and less energy are used than that of conventional latex paints. The proof for the durability of biopaints is the old and historical art works which lasts even today. The notable characteristics of biopaints aere low odour, excellent durability and washable finish. The main objective of the study is to formulate biopaint using the dye extracted from *Anisochilous carnosus* (L.f.) Wall as pigment and to test its effect against algal growth in the wall of a temple in Erode, Tamil Nadu, India.

Chapter 2: Materials used and the Methodology

Biopaint formulation

Collection of plant materials

Anisochilus carnosus (L.f.) Wall was collected from Vinayaka mission's 1008 Sivalayakailayam which is situated in Ariyanoor, Salem district. It is situated between 78°3'E latitude and 11°8' N longitude. The maximum and minimum temperature of this area are 35.3°c and 20.4°c respectively. The relative humidity of the area fluctuates from 74 % to 85 %. The place is located at an altitude of 279 m. The plant was identified with the help of floras the Flora of Presidency of Madras (Gamble and Fischer, 1935)and Flora of Tamil Nadu Carnatic (Mathew, 1983).

Extraction of dye

The leaves were collected from the plant and washed thoroughly 2-3 times with running tap water. The leaf materials were then air dried in shade just to remove the water particles. The leaves from the plant source were crushed and dissolved in distilled water and allowed to boil in a beaker kept over water bath for quick extraction for 2 hours. The dye was prepared in different concentrations namely 5%, 10%, 25%, 50% and 75%. 5% dye was obtained by boiling 5g leaves in 100 ml water. 10% dye was obtained by boiling 10g leaves in 100 ml of water. 25% dye was obtained by boiling 75g of leaves in 300ml water. 50% dye was obtained by boiling 250g of leaves in 500ml water and 75% was obtained by boiling 563g of leaves in 750ml water. The dye was completely extracted from the leaves by the end of 2 hours. The solutions were filtered for immediate use.

Preparation of Biopaint

The dye extracted from *Anisochilus carnosus* (L.f.)Wall was used as the pigment or the colour source. Starch (from flour) is the best natural binder for biopaint. It keeps the dye glued to the surface.

Limestone powder is used as fillers. It creates texture and add bulk to the paint. This is mainly added to achieve a workable consistency. In biopaint water is used as a solvent.

Lynn Edward and Julie Lawler, 2003, formulated the steps for preparation of biopaint. The steps are as follows:

- Starch (50g) is dissolved in boiling water (250 ml). This solution is used as a binder.
- The dye (250 ml) from the plant is extracted by boiling the leaves (60 g) in water (250 ml) followed by filtration.
- To the filtrate add the binder and stir well.
- Limestone powder (1500 g) is taken as filler and added to the above mixture to achieve the desired consistency as well as quantity.

Different concentrations of Biopaint

Different concentrations of biopaint such as 5%, 10%, 25%, 50% and 75% were made to test the efficiency, (Lynn Edwards and Julia Lawler, 2003).

5% Biopaint

10 g of starch is dissolved in 60 ml of distilled water by boiling. This solution acts as binder. 50 ml of 5% dye is dissolved in this binder. 1500g of limestone powder is added to this mixture. The paintable consistency is obtained by adding water as a solvent (1390 ml).

10% Biopaint

25 g of starch is dissolved in 125 ml of distilled water by boiling. This solution acts as binder. 100 ml of 10% dye is dissolved in this binder. 1500g of limestone powder is added to this mixture. The paintable consistency is obtained by adding water as a solvent (1275 ml).

25% Biopaint

50 g of starch is dissolved in 250 ml of distilled water by boiling. This solution acts as binder. 250 ml of 25% dye is dissolved in this binder. 1500g of limestone powder is added to this mixture. The paintable consistency is obtained by adding water as a solvent (1000 ml).

50% Biopaint

100 g of starch is dissolved in 500 ml of distilled water by boiling. This solution acts as binder. 500 ml of 50% dye is dissolved in this binder. 1500g of limestone powder is added to this mixture. The paintable consistency is obtained by adding water as a solvent (500 ml).

75% Biopaint

150 g of starch is dissolved in 750 ml of distilled water by boiling. This solution acts as binder. 750 ml of 5% dye is dissolved in this binder. 1500g of limestone powder is added to this mixture. The solvent is avoided in this case.

The bio paint is ready for application.

Biopaint -application

The wall subjected to bio painting was first cleaned to remove dust and algae before applying the bio paint.

Square plots and partition assumptions

Squares of 100cm×100cm or 1m×1m were marked on the temple wall. The potentiality of the applied paint could be calculated by assuming 100 small squares of 10cm×10cm each within the large one. These assumptions were made only after drying of the applied paint. Permanent marker was used for drawing squares over the coated area. This is done to test the degrading and the resistance potential of the various concentrations of biopaint at different coating levels. This could be done effectively by taking into account the colour fading of the paint and the biofilm formation in each small assumed squares (10cm×10cm). The potential was calculated by subtracting the damaged boxes from the undamaged ones.the area damaged could be easily calculated using the formula Area = $(side)^2$. This counting was done separately for each square plots and at regular time intervals (10th, 15th, 20th, 25th, 30th, 35th, 40th and 45th days). The values so obtained are converted to percentage and are tabulated.

Coatings of biopaint

For each concentration of prepared biopaint, single, double and triple coatings were applied separately on three square plots. Accordingly, as there are 5 different concentrations of biopaint, 15 square plots (each 100cm×100cm) were taken for the present study. In the case of single coating a single layer of paint was applied over the selected area and was left to dry overnight and subjected to further examinations. In double coatings, first a single layer was applied and it was left to dry overnight and on the next day another layer of paint was applied over the first one. To apply triple coats the same procedure is followed but a three layers of paints was applied over the selected area by leaving drying gaps between each coat.

Meteorological data

Meteorological data of Bhavani taluk of Erode district was collected from Tamil Nadu Agriculture weather report, a website of Tamil Nadu Agricultural University (TNAU). It included data regarding air temperature, relative humidity, wind speed, rainfall, solar radiation and atmospheric pressure (*tnau.ac.in*)

Chapter 3: Results

Botanical studies of the dye yielding plant

Plant identification

The study plant was identified with the help of standard flora such as 'Flora of Presidency of Madras' (Gamble, 1935) and also with the help of 'The Flora of the Tamilnadu Carnatic India' (Mathew, 1983).

Systematic position

Class	:	Dicotyledons
Subclass	:	Gamopetalae
Series	:	Bicarpellatae
Cohort	:	Lamiales
Family	:	Lamiaceae
Genus	:	*Anisochilus*
Species	:	*carnosus* (L.f.)Wall

An erect herb with tetragonous stems; flowers in long peduncled spikes; small in flowers but large strobilate in fruits and covered with red glands; corolla pale purple; nutlets orbicular; compressed; shining and brown (Gamble,1935)

Testing the effect or potential of the various concentrations of Biopaint

Biopaint of concentrations 5%, 10%, 25%, 50% and 75% were prepared. For each concentration of biopaint, 3 different coatings were made(single, double and triple coats).The efficiency of bio paint was tested for 45 days from July 2 to August 15, 2016. Biopaint examination were carried out at regular time intervals on $10^{th}, 15^{th}, 20^{th}, 25^{th}, 30^{th}, 35^{th}$ and 45th day of the study period.

Effect of 5% Biopaint

The biopaint was applied in three $1m^2$ area as single, double and triple coats.

Single coating

In this case, the $1m^2$ square box on the temple wall was coated with single layer of biopaint at 5 am in the morning on July 2. The first examination was carried out on the 10th day and the 2nd examination was made on the 15th day. On that day out of the total area (10000 cm^2) 500cm^2 was found to be affected by colour fading. Colour fading was due to different weather conditions. This increased gradually on the successive days and at the end of the final day(45th day), a total area of 5300 cm^2 was found to be affected by fading and formation of biofilm. The remaining 4700 cm^2 was left unaffected. The algal growth started from the 22nd day. Along with these, the degrading as well as the resistance percentage were also calculated. The single coat of 5% concentration of bio paint exhibited a degrading percentage which increased from 5% on the 15th day to 53% at the end of the 45th day. In other words, the paint exhibited a resistance percentage which decreased from 100% on the 10th day to 47% on the 45th day. The average degrading percentage was 26.87% and the average resistance percentage was 73.12%.

Double coating

Here the $1m^2$ square area was coated with a single layer of paint on the July 2 at 5 am. It was left to dry overnight and on July 3 over the dried layer another layer of coating was given at 5 am. The first examination was carried out on the 10th day but the paint remained unaffected. The second examination was made on the 15th day and of the total area (10000 cm^2)400 cm^2 was found affected by

9 | Biopaint- Preparation and Testing

external factors which resulted in colour fading. This gradually increased on the successive days and at the end of the final day (45th day), a total area of 4800 cm^2 was found to be affected either by colour fading or by formation of biofilm and the remaining 5200 cm^2 was left unaffected. The algal growth started from the 23rd day. The degrading and the resistance percentage were calculated using the above results. The double coats of 5% concentration of biopaint exhibited a degrading percentage which increased from 4% on the 15th day to 48% at the end of the 45th day. The paint exhibited a resistance percentage which decreased from 100% on the 10th day to 52% on the 45th day. The average degrading percentage was 21.5% and the average resistance percentage was 78.5%.

Triple coating

In the case of triple coats, the 1m^2 square wall was coated with a single layer of paint on July 2 at 5 am and left to dry overnight and on July 3 another coat was applied over the first one at 5 am. In the same way, the third coating was also given on July 4 at 5 am. The paint remained unaffected for the first 15 days but on the 20th day out of the total area (10000 cm^2) 400 cm^2 was found to be affected which was indicated by colour fading. At the end of 45th day, a total area of 3900 cm^2 was affected by both colour fading and algal biofilm formation and the remaining 6100 cm^2 remained as such without any change. The algal growth started from 24th day. The triple coats showed a degrading percentage which increased from 4% on the 20th day to 39% on the 45th day and had a resistance percentage which decreased from 100% on the 10th day to 61% on the 45th day. 17.12% and 82.87% were the average degrading and average resistance percentage respectively.

Effect of 10% Biopaint

The biopaint was applied in three 1m^2 area as single, double and triple coats.

Single coating

In this case, the 1m² square box on the temple wall was coated with single layer of biopaint at 5.10 am in the morning on July 2. On the 15th day 300 cm² out of 10000 cm² was affected by external factors which faded its colour. At the end of 45th day, a total area of 5100 cm² was found to be affected but both colour fading and the formation of biofilm, the remaining 4900 cm² was left unaffected. The algal growth started from 28th day. While calculating the degrading and the resistance percentage, the following results were obtained. The single coat (10%) had a degrading percentage which increased from 3% on the 15th day to 51% on the 45th day. Whereas it had a resistance percentage which decreased from 100% on the 10th day to 49% on the 45th day. The average degrading and resistance percentage were 21.87% and 78.12% respectively.

Double coating

Here the 1m² square area was coated with a single layer of paint on the July 2 at 5.10 am. It was left to dry overnight and on July 3 over the dried layer another layer of coating was given at 5 am. During first two examinations (10th & 15th day) no considerable changes were noted. But on the 20th day an area of 300 cm² was found to be affected by external agents. At the end of final day (45th day) an area of 3700 cm² was found to be affected by colour fading and bio film formation. The rest 6300 cm² was left unchanged. The algal growth started from the 30th day. It was noted that the degrading percentage of this concentration increased from 4% on the 20th day 7% on the 45th day. The resistance percentage decreased from 100% on the 10th day to 63% on the 45th day. The average degrading percentage was 16.12% and the average resistance percentage was 83.87%.

Triple coating

In the case of triple coats, the 1m² square wall was coated with a single layer of paint on July 2 at 5.10 am and left to dry overnight and on July 3 another coat was applied over the first one at 5 am. In the same way, the third coating was also given on July 4 at 5 am. The paint did withstand from the attack of external factors and bio films for the first 20 days. But on the 25th day an area of 300 cm^2 was found to be affected by external agents. By the end of 45th day, an area of 3100 cm^2 out of 10000 cm^2 was affected resulting in colour fading and bio film formation whereas the rest 6900 cm^2 remained without any change. The algal growth started from the 30th day. On calculating the degrading and resistance percentage, it was noted that the degrading percentage increased from 3% on the 25th day to 31% on the 45th day. The resistance percentage was decreased from 100% on the 10th day to 69% on the 45th day. The average degrading percentage was 10.75% and the average resistance percentage was 89.25%.

Effect of 25% Biopaint

The biopaint was applied in three 1m^2 area as single, double and triple coats.

Single coating

In this case, the 1m^2 square box on the temple wall was coated with single layer of biopaint at 5.15 am in the morning on July 2. The biopaint exhibited changes only on the 20th day. On this day an area of 300cm^2 has shown fading colour. On the 45th day an area of 4700 cm^2 was affected by both colour fading and bio films formation whereas the rest 5300 cm^2 remained as such. The algal growth started from 29th day. The degrading percentage was found to be increased from 3% on to 20th day to 47% on the 45th day. The resistance percentage was found to be decreased from 100% on the 10th day to 53% on the 45th day. 20.25% and 79.75% were the average and resistance percentages respectively.

Double coating

Here the $1m^2$ square area was coated with a single layer of paint on the July 2 at 5.15 am. It was left to dry overnight and on July 3 over the dried layer another layer of coating was given at 5 am. The paint resisted for 15 days. On 20th day it was found that 200 cm^2 started to fade in colour. By the end of 45th day an area of 3500 cm^2 was affected and the rest 6500 cm^2 remained unaffected. The algal growth started from 33rd day. Using the above results, the resistance and degrading percentage of the bio paint were calculated. The degrading percentage increased from 2% on the 20th day to 35% on the 45th day and the resistance percentage decreased from 100% on the 10th day to 65% on the 45th day. The average degrading percentage was 14% and the average resistance percentage was 86%.

Triple coating

In the case of triple coats, the $1m^2$ square wall was coated with a single layer of paint on July 2 at 5.15 am and left to dry overnight and on July 3 another coat was applied over the first one at 5 am. In the same way, the third coating was also given on July 4 at 5 am. The paint did withstand from the activities of external agents and bio films for the 20 days. But on the 25th day in area of 300 cm^2 was affected resulting in colour fading. A total area of 3300cm^2 was affected by both external agents and biofilms at the end of 45th day 6700 cm^2 area was unaffected. The algal growth started from 34th day. The degrading percentage increased from 3% on the 25th day to 33% on the 45th day and the resistance percentage decreased from 100% on the 19th day to 67% on the 45th day. The average degrading and resistance percentage were 12.12% and 87.87% respectively.

Effect of 50% Biopaint

The biopaint was applied in three $1m^2$ area as single, double and triple coats.

13 | Biopaint- Preparation and Testing

Single coating

In this case, the 1m² square box on the temple wall was coated with single layer of biopaint at 5.20 am in the morning on July 2. No considerable change was encountered for the first 15 days but on the 20th day an area of 100 cm² exhibited colour fading. By the end of 45th day an area of 3700 cm² was affected and shown colour fading and biofilm formation whereas the rest 6300 cm² remained as such. The algal growth started from 33rd day. The degrading percentage was found to be increased from 1% on the 20th day to 37% on the 45th day. The resistance percentage was decreased from 100% on the 10th day to 63% on the 45th day. The average degrading percentage was 14.5% and the average resistance percentage was 85.5%.

Double coating

Here the 1m² square area was coated with a single layer of paint on the July 2 at 5.20 am. It was left to dry overnight and on July 3 over the dried layer another layer of coating was given at 5 am. The bio paint did withstand for first 15 days and on the 20th day 200cm² was affected resulting in colour fading. On the final day of the total 10000 cm² an area of 2500 cm² was affected exhibiting bio film formation and colour fading and the rest 7500 cm² was left unchanged. The algal growth started from 35th day. The degrading percentage was increased from 2% on the 20th day to 25% on the 45th day. The resistance percentage was decreased from 100% on the 10th day to 75% on the 45th day. The average degrading percentage was 9.62% and the average resistance percentage was 90.37%.

Triple coating

In the case of triple coats, the 1m² square wall was coated with a single layer of paint on July 2 at 5.20 am and left to dry overnight and on July 3 another coat was applied over the first one at 5 am. In the same way, the third coating was also given on July 4 at 5 am. It remained unaffected for the first 20 days. On the 25th day an area of

100 cm^2 exhibited colour fading. By the end of 45th day a total area of 1700 cm^2 was affected and the rest 8300 cm^2 was left unaffected. The algal growth started from 35th day. But the bio film formation was less when compared to the previous concentration. With this data the degrading and resistance percentage were calculated. The degrading percentage increased from 1% on the 25th day to 17% on the 45th day. In the same way, the resistance percentage decreased from 100% on the 10th day to 83% on the 45th day. The average degrading percentage was 5.75% and the average resistance percentage was 94.25%.

Effect of 75% Biopaint

The biopaint was applied in three 1m^2 area as single, double and triple coats.

Single coating

In this case, the 1m^2 square box on the temple wall was coated with single layer of biopaint at 5.30 am in the morning on July 2. The paint remained as such for first 15 days on the 20th day 100 cm^2 area was affected. On the 45th day it was noted that only 2500cm^2 area was affected and the rest 7500 cm^2 remained as such. The algal growth started from 35th day. The degrading percentage was found to be increased from 1% on the 20th day to 25% on the 45th day. The resistance percentage was decreased from 100% on the 10th day to 75% on the 45th day. The average degrading percentage was 10.37% and the average resistance percentage was 89.62%.

Double coating

Here the 1m^2 square area was coated with a single layer of paint on the July 2 at 5.30 am. It was left to dry overnight and on July 3 over the dried layer another layer of coating was given at 5 am. The paint remained unaltered till the 20th day. On the 25th day 300 cm^2 area was affected. By the end of 45th day an area of 2100 cm^2 was affected and the rest 7900 cm^2 was unaffected. The algal growth started from

35th day. The degrading percentage increased from 3% on the 25th day to 21% on the 45th day. The resistance percentage was decreased from 100% on the 10th day to 79% on the 45th day. The average degrading percentage was 8.5% and the average resistance percentage was 91.5%.

Triple coating

In the case of triple coats, the 1m^2 square wall was coated with a single layer of paint on July 2 at 5.30 am and left to dry overnight and on July 3 another coat was applied over the first one at 5 am. In the same way, the third coating was also given on July 4 at 5 am. The first three examinations encountered no change but on the 25th day 100 cm^2 area was affected. By the end of 45th day an area of 1200 cm^2 was affected whereas the rest 8800 cm^2 was left unaffected. The algal growth started from 35th day. But the bio film formation was less when compared to the previous concentration. The degrading percentage increased from 1% on the 25th day to 12% on the 45th day and the resistance percentage was decreased from 100% on the 10th day to 88% on the 45th day. The average degrading percentage was 4.12% and the average resistance percentage was 95.87%

Algal species colonizing after the Biopaint application

The algal biofilms developed on the surface of the biopaint was collected using scalpels and carried to laboratory in polythene bags. In the lab, the slides were prepared for sample. The slides then were examined using low and high power pre calibrated Olympus microscope using natural daylight. From the observation seven species were recorded Of these 2 are colonial forms, 1 unicellular forms and four filamentous forms. The species are *Chroococcus minor, Aphanocapsa grevelli, Gloeocapsa atrata, Oscillatoria limosa, Oscillatoria willei, Lyngbya major* and *Lyngbya digueiti* were recorded.

Meteorological data of the study area

The meteorological data of the study area, Bhavani taluk of Erode district was recorded from the website of Tamil Nadu Agricultural University. It included details like temperature, relative humidity, rainfall, atmospheric pressure, solar radiation and wind speed of the study area. The study area experienced rainfall for 7 days during the study period (July 2 – August 15).the maximum rainfall was recorded on July 26 (55.5 mm), followed by 34.5 mm on July 29, 12 mm on July 17, 6.5 mm on July 21, 5.5 mm on August 14, 3 mm on July 18 and 2.5 mm on July 28. The rainfall started on July 17 and after 6 days from that the algal biofilm started to appear on the biopaint coated wall. Single coating was the first to get affected followed by double coats and the least affected was triple coats. Among the concentrations, 5% was most affected and 75% was least affected. The algal biofilm formation decreased with increase in concentration of biopaint.

Figure: Example for the method used for the study

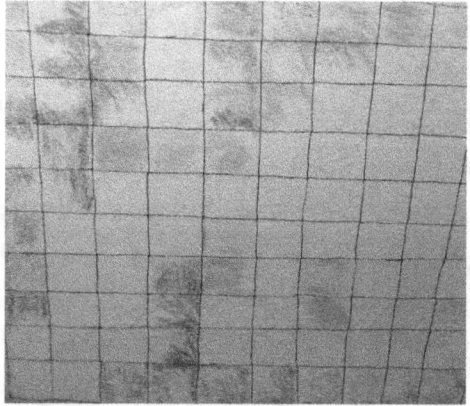

Table 1 – Overall degrading and resistant percentage of coatings at different concentrations of biopaint at various coatings

Concentration of Biopaint	Single Coating		Double Coatings		Triple Coatings	
	Degrading %	Resistant %	Degrading %	Resistant %	Degrading %	Resistant %
5% Conc.	26.87%	73.12%	21.5%	78.5%	17.12%	82.87%
10% Conc.	21.87%	78.12%	16.12%	83.87%	10.75%	89.25%
25% Conc.	20.25%	79.75%	14%	86%	12.12%	87.87%
50% Conc.	14.5%	85.5%	9.62%	90.37%	5.75%	94.25%
75% Conc.	10.37%	89.62%	8.5%	91.5%	4.12%	95.87%

Table 2- Algal appearance on wall at different coatings of various concentrations of biopaint and the rainfall recorded.

Date	5%			10%			25%			50%			75%			Rainfall
	S	D	T	S	D	T	S	D	T	S	D	T	S	D	T	
July 2	-	-	-	-	-	-	-	-	-	-	-	-	-	-	-	
July 3	-	-	-	-	-	-	-	-	-	-	-	-	-	-	-	
July 4	-	-	-	-	-	-	-	-	-	-	-	-	-	-	-	
July 5	-	-	-	-	-	-	-	-	-	-	-	-	-	-	-	
July 6	-	-	-	-	-	-	-	-	-	-	-	-	-	-	-	
July 7	-	-	-	-	-	-	-	-	-	-	-	-	-	-	-	
July 8	-	-	-	-	-	-	-	-	-	-	-	-	-	-	-	
July 9	-	-	-	-	-	-	-	-	-	-	-	-	-	-	-	
July 10	-	-	-	-	-	-	-	-	-	-	-	-	-	-	-	
July 11	-	-	-	-	-	-	-	-	-	-	-	-	-	-	-	
July 12	-	-	-	-	-	-	-	-	-	-	-	-	-	-	-	
July 13	-	-	-	-	-	-	-	-	-	-	-	-	-	-	-	
July 14	-	-	-	-	-	-	-	-	-	-	-	-	-	-	-	
July 15	-	-	-	-	-	-	-	-	-	-	-	-	-	-	-	
July 16	-	-	-	-	-	-	-	-	-	-	-	-	-	-	-	
July 17	-	-	-	-	-	-	-	-	-	-	-	-	-	-	-	12 mm
July 18	-	-	-	-	-	-	-	-	-	-	-	-	-	-	-	3 mm
July 19	-	-	-	-	-	-	-	-	-	-	-	-	-	-	-	
July 20	-	-	-	-	-	-	-	-	-	-	-	-	-	-	-	
July 21	-	-	-	-	-	-	-	-	-	-	-	-	-	-	-	6.5 mm
July 22	-	-	-	-	-	-	-	-	-	-	-	-	-	-	-	
July 23	+	-	-	-	-	-	-	-	-	-	-	-	-	-	-	
July 24	+	+	-	-	-	-	-	-	-	-	-	-	-	-	-	
July 25	+	+	+	-	-	-	-	-	-	-	-	-	-	-	-	
July 26	+	+	+	-	-	-	-	-	-	-	-	-	-	-	-	55.5 mm
July 27	+	+	+	-	-	-	-	-	-	-	-	-	-	-	-	

Biopaint- Preparation and Testing

Date															
July 28	+	+	+	-	-	-	-	-	-	-	-	-	-	-	2.5 mm
July 29	+	+	+	+	-	-	-	-	-	-	-	-	-	-	34.5 mm
July 30	+	+	+	+	-	-	+	-	-	-	-	-	-	-	
July 31	+	+	+	+	+	+	+	-	-	-	-	-	-	-	
Aug 1	+	+	+	+	+	+	+	-	-	-	-	-	-	-	
Aug 2	+	+	+	+	+	+	+	-	-	-	-	-	-	-	
Aug 3	+	+	+	+	+	+	+	+	-	+	-	-	-	-	
Aug 4	+	+	+	+	+	+	+	+	+	+	-	-	-	-	
Aug 5	+	+	+	+	+	+	+	+	+	+	+	+	+	+	
Aug 6	+	+	+	+	+	+	+	+	+	+	+	+	+	+	
Aug 7	+	+	+	+	+	+	+	+	+	+	+	+	+	+	
Aug 8	+	+	+	+	+	+	+	+	+	+	+	+	+	+	
Aug 9	+	+	+	+	+	+	+	+	+	+	+	+	+	+	
Aug 10	+	+	+	+	+	+	+	+	+	+	+	+	+	+	
Aug 11	+	+	+	+	+	+	+	+	+	+	+	+	+	+	
Aug 12	+	+	+	+	+	+	+	+	+	+	+	+	+	+	
Aug 13	+	+	+	+	+	+	+	+	+	+	+	+	+	+	
Aug 14	+	+	+	+	+	+	+	+	+	+	+	+	+	+	5.5 mm
Aug 15	+	+	+	+	+	+	+	+	+	+	+	+	+	+	

+ Presence of Algae

- Absence of algae

S- Single coat

D- Double coats

T- Triple coats

Table 3- Meteorological data of the study area

Date	Air temperature(°C) Max	Min	Relative Humidity(%)	Wind speed (Kmph)	Rainfall (mm)	Solar Radiation (cal/cm^2)	Atmospheric Pressure (hpa)
July 2	36.9	24.3	52.7	12.1	0.0	639.9	982.6
July 3	36.7	25.4	50.1	11.2	0.0	700.8	983.3
July 4	37.1	25.2	57.2	7.3	0.0	664.3	984.2
July 5	34.0	24.6	67.5	6.8	0.0	567.4	984.8
July 6	34.2	24.9	66.0	6.1	0.0	542.8	984.1
July 7	34.1	25.6	66.2	7.7	0.0	569.5	984.0
July 8	34.1	24.7	64.6	6.2	0.0	529.6	983.8
July 9	35.5	25.1	60.1	8.3	0.0	511.8	983.1
July 10	35.6	26.1	56.2	8.4	0.0	573.2	983.0
July 11	34.9	26.2	56.9	9.3	0.0	469.3	982.7
July 12	35.5	26.5	54.5	9.9	0.0	586.6	982.9
July 13	35.8	25.0	52.3	9.0	0.0	661.0	984.1
July 14	35.4	24.4	61.0	7.8	0.0	630.5	983.1
July 15	37.3	25.5	59.1	7.0	0.0	686.1	982.8
July 16	36.4	26.1	61.1	6.5	0.0	621.7	983.5
July 17	36.3	24.7	74.3	5.0	**12.0**	609.4	985.2
July 18	34.0	24.6	82.4	3.8	**3.0**	528.1	986.8
July 19	33.9	24.3	73.5	5.0	0.0	568.3	986.4
July 20	34.6	25.8	61.6	5.8	0.0	602.3	985.6
July 21	35.3	24.5	70.5	5.1	**6.5**	587.2	984.6
July 22	32.8	25.0	62.6	5.8	0.0	515.5	984.0
July 23	33.7	24.3	63.1	5.1	0.0	494.7	983.6
July 24	33.5	23.9	67.7	5.0	0.0	533.0	983.3

July 25	35.0	25.4	69.2	4.7	0.0	515.8	983.1
July 26	35.9	23.3	77.9	5.7	**55.5**	584.4	983.4
July 27	32.2	24.6	79.4	3.7	0.0	552.2	983.4
July 28	33.7	24.5	83.2	4.0	**2.5**	541.3	984.1
July 29	23.8	23.7	80.7	5.0	**34.5**	658.5	983.7
July 30	32.8	23.8	77.6	3.5	0.0	559.2	984.0
July 31	31.7	24.0	80.5	4.7	0.0	524.8	985.2
Aug 1	34.1	23.3	66.4	5.1	0.0	607.0	985.3
Aug 2	34.3	22.8	64.4	6.4	0.0	610.7	984.4
Aug 3	33.7	23.1	60.4	8.6	0.0	582.5	982.6
Aug 4	35.4	24.3	54.9	7.9	0.0	696.6	982.4
Aug 5	34.3	25.4	55.7	8.1	0.0	541.6	983.4
Aug 6	34.2	23.7	60.0	7.6	0.0	506.6	984.4
Aug 7	36.6	24.0	59.4	7.2	0.0	719.8	984.4
Aug 8	35.4	23.8	61.0	5.8	0.0	573.7	985.6
Aug 9	35.7	25.4	59.5	6.2	0.0	662.9	985.5
Aug 10	36.6	24.3	59.5	6.2	0.0	646.6	985.3
Aug 11	35.3	23.7	60.0	6.6	0.0	579.1	985.8
Aug 12	36.5	24.3	59.2	5.9	0.0	674.8	986.3
Aug 13	36.9	25.8	58.6	7.8	0.0	672.3	985.6
Aug 14	36.8	25.0	64.9	6.1	**5.5**	613.5	984.8
Aug 15	34.1	24.4	63.0	6.8	0.0	554.6	984.8

Chapter 4: Discussion

The colonization of biofilms cause the unacceptable appearance of staining on stone surface by biogenic pigments and the production of extracellular polymeric substances (EPS) that cause mechanical stress to the mineral structure due to shrinking and swelling cycles. This could also forms the growth medium for the deteriogenic fungi which cause further degradation. The early presence of biofilms on the exposed stone surfaces accelerates the accumulation of atmospheric pollutants (Crispim, 2005). This study is the first report on the production and use of biopaint for the control of biofilm on wall.

The early reports on biopaint preparation (Nithyanandha sastry *et al.,* 2016 and Mohibah Musa *et al.*, 2013) involved the preparation of biopaint using fungal strains and methyl ester from palm oil respectively as colouring agents. In contrast to this the present study involved the use of dye extracted from *Anisochilus carnosus* (L.f) Wall as the colouring agent in the biopaint. This is an exclusive work of using a light coloured plant dye in the preparation of biopaint.

Nithyanandha sastry *et al.*, 2016 prepared natural lime, milk, oil-in-water emulsion paints using natural microbial colours and ecofriendly ingredients. They also carried out biopaint application and evaluated them. Similarly the present study involved the preparation of biopaint with ecofriendly ingredients like plant dye, starch as binder, water and limestone powder. The current study also involved the biopaint application and evaluation.

Early reports on biopaint focused on preparation of a single concentration of biopaint (Mohibah Musa *et al.,* 2013). In contrast to this the present study involved preparation of biopaint of five different concentrations like 5%, 10%, 25%, 50% and 75% and examining the potentiality of each concentration by applying them as single, double and triple coats over the selected wall surface. This is uniqueness to the present study.

The previous studies did not record the meteorological data of the study area. Hence this study forms the first to record the meteorological data of study area to understand the parameters responsible for the growth of algal biofilm. It was recorded that the study area experienced rainfall for 7 days during the study period. The rainfall started on July 17 and after 6 days from that the algal biofilm started to appear on the coated biopaints. Single coating was the first to get affected followed by double coats and the least affected was triple coats. Among the concentrations, 5% was most affected and 75% was least affected.

Altogether the purpose of the study was to document the algal flora on the temple wall, temple tank, temple floor and temple statue of three temples of Erode district and to control its growth through naturally prepared ecofriendly biopaint. This study may help in future to determine appropriate actions for the prevention of algal growth on the exposed walls of temples and monuments and to protect them from biodeterioration.

Chapter 5: Summary and Conclusion

- The bio paint was prepared in five different concentrations of plant dye extracted from *Anisochilus carnosus* (L.f.)Wall (5%, 10%, 25%, 50% and 75%) along with dissolved starch (binder), water which acts as a solvent and limestone powder as filler.
- For each concentration of biopaint three different coatings (single, double and triple coats) were applied to the temple wall flora.
- The temple wall subjected to testing was divided into 15 small squares of 1m^2 each for separately applying various coats of different concentrations of bio paints.
- The efficiency of bio paint was tested for 45 days from July 2 to August 15, 2016.
- The wall painted with single, double and triple coats of 5% concentration of biopaint showed deterioration areas of 5300cm^2, 4800cm^2 and 3900cm^2 respectively.
- The wall painted with single, double and triple coats of 10% concentration of biopaint showed deterioration areas of 5100cm^2, 3700cm^2 and 3100cm^2 respectively.
- The wall painted with single, double and triple coats of 25% concentration of biopaint showed deterioration areas of 4700cm^2, 3500cm^2 and 3300cm^2 respectively.
- The wall painted with single, double and triple coats of 50% concentration of biopaint showed deterioration areas of 3700cm^2, 2500cm^2 and 1700cm^2 respectively.
- The wall painted with single, double and triple coats of 75% concentration of biopaint showed deterioration areas of 2500cm^2, 2100cm^2 and 1200cm^2 respectively.
- As per the meteorological data rainfall started on July 17 and after 6 days from that the algal biofilm started to appear on the coated biopaints. Single coating was the first to get affected followed by double coats and the least affected was triple coats. Among the concentrations, 5% was most affected and 75% was least affected.

- The increase in concentration of the dye pigment and coatings controlled the algal growth to some extent.
- The genera that reported after biopaint application were *Chroococcus*, *Gloeocapsa*, *Aphanocapsa*, *Oscillatoria* and *Lyngbya*.
- Thus the biopaint from the dye of *Anisochilus carnosus* (L.f.)Wall prove to control algal growth to some extent.
- This study is the first report on the preparation and use of plant based biopaint for the control of algal growth on wall through naturally prepared ecofriendly biopaint.
- This study may help in future to take suitable actions for the prevention of biofilm on the exposed monuments and to protect them from microbial degradation.

Bibliography

Anisha B. Khan and G. Kulathuran (2010).Composition of microorganisms in deterioration of stone structures and monuments. *The Bioscan*.1:57-67.

Cezar A. Crispim, Peter M. Gaylarde and Christine C. Gaylarde (2003). Algal and cyanobacterial biofilms on calcareous historic buildings. *Current microbiology*. 46:79-82.

Divya Lekshmi R.B. and D. Ravi (2013). Extraction of natural dyes from selected plant sources and its application in fabrics. *International Journal of Textile and fashion Technology*. 3(2):53-60.

Gamble J.S. and C.E.C. Fischer (1915-1935). Flora of Presidency of Madras. *London*. Rep.ed. 1957.

Gaylarde C.C. and P.M. Gaylarde (2002). Biodeterioration of historic buildings in Latin America. *DBMC*. 171:1-9.

http://en.m.wikipedia.org/wiki/paint.

http://tawn.tnau.ac.in

Jothi D. (2008). Extraction of natural dyes from African marigold flower (*Tagetes erecta* L.) for textile colouration. *Autex Research Journal*. 8(2):49-53.

Keka Sinha, Papita Darsaha and Siddhartha datta (2012). Extraction of natural dye from petals of flame of the forest (*Butea monosperma*) flower:Process optimization using response surface methodology (RSM). *Dyes and pigments*. 94(2):212-216.

Kulkarni S.S., A.V.Gokhale, U.M. Bodake and G.R. Pathade (2011). Cotton dyeing with natural dye extracted from Pomegranate (*Punica granatum*) peel. *Universal journal of Environmental Research and Technology*. 1(2):135-139.

Mathew K.M. (1983). The Flora of Tamil Nadu Carnatic. *Rapinet Herbarium, Trichirapalli*, 1-3: 399

Mohibah Musa, Miradatul Najwa Muhd Rodhi, Najmiddin Yaakob, Ku Halim Ku Hamid and Juferi Idris (2013). Development of Bio-based paint by using methyl esters from Palm oil for corrosion inhibitor. *The Malaysian Journal of Analytical sciences.* 17(1):30-37.

Nithyananda Sastry D, T. Prabhakar, M. Lakshmi Narasu (2016). Studies on preparation of biopaints using fungal biocolors. *Pigments and Resin Technology.* 45(2):79-85.

www.ingramcontent.com/pod-product-compliance
Lightning Source LLC
Chambersburg PA
CBHW030040230526
45472CB00002B/610